Jaildo Assis

Difficulties in understanding and using rational numbers:

Jaildo Assis

Difficulties in understanding and using rational numbers:

An evaluative perspective in youth and adult education

ScienciaScripts

Imprint

Any brand names and product names mentioned in this book are subject to trademark, brand or patent protection and are trademarks or registered trademarks of their respective holders. The use of brand names, product names, common names, trade names, product descriptions etc. even without a particular marking in this work is in no way to be construed to mean that such names may be regarded as unrestricted in respect of trademark and brand protection legislation and could thus be used by anyone.

Cover image: www.ingimage.com

This book is a translation from the original published under ISBN 978-613-9-62316-7.

Publisher:
Sciencia Scripts
is a trademark of
Dodo Books Indian Ocean Ltd. and OmniScriptum S.R.L publishing group

120 High Road, East Finchley, London, N2 9ED, United Kingdom
Str. Armeneasca 28/1, office 1, Chisinau MD-2012, Republic of Moldova, Europe
Printed at: see last page
ISBN: 978-620-7-72125-2

SUMMARY

The secret is not to chase the butterflies... It's taking care of the garden so that they come to you. Mârio Quintana.

ACKNOWLEDGMENTS

To all those who seek to investigate mathematics in order to find more accessible paths to development in general.

To those who contributed to the completion of this work; to those who are waiting for its successful conclusion.

The teacher for her patience and encouragement in carrying out the research; The family for their unconditional support, always encouraging and wishing for the best result.

SUMMARY

This work is concerned with the development of science and technology, and aims to investigate how we can help minimize the many situations that interfere with the development of fractional activities in EJA - Youth and Adult Education, and to research the development of strategies for working with tools that can act by enabling and making EJAI students feel more comfortable with rational numbers. In terms of methodology, 32 students from two high school classes, of both sexes, belonging to two state schools in the state of Pernambuco, were interviewed. The students answered a questionnaire, interview and test, which included the following items: Comparison of number of students and classes, comparison of number of students gender/age, number of students and liking for the subject, comparison of number of students and fractions, comparison of number of students and repetition rate, comparison of correct answers per class, difference between correct answers and gender. As far as the results are concerned, this study showed that the elements introduced to reduce difficulties in relation to rationals were of great significance in teaching and learning, especially from the point of view of the evolution between the first and second tests given to the students. It was possible to evaluate the practice by comparing results in relation to the difficulties we tried to minimize. The conclusions are: it is necessary for EJA students to understand rational numbers; through strategies that allow them to perceive the registers of semiotic representations from an evaluative perspective, not only in mathematics, but in various areas of knowledge.

Keywords: Rational numbers. EJA. Evaluation.

1   **INTRODUCTION**

Unfortunately, students who are outside the age bracket enter Youth and Adult Education (EJA) in the hope of developing skills that will enable them to enter the job market, but they are faced with teachers who have not been prepared to deal with them in a way that is appropriate to their reality.

The teachers are the same as those who teach high school, and therefore practice the same methodology applied in conventional education. In this way, students naturally tend to drop out, because the "stigmas" often heard by the math teacher soon emerge. These are expressions like: "I hate math", "I don't know math", "I never learned math", etc. As an educator, you have to look for efficient methodologies for teaching and learning mathematical content, using the connection between practical and systematic knowledge, which promotes inclusion.

Welcoming these people is fundamental, because they are underprivileged people who find themselves on the margins in the hope of a future and of finding themselves again, but public policies have not developed practices that promote raising their self-esteem, and no thought has been given to the intellectual level of these citizens. Both students and teachers need to be respected, and in both situations there has been no efficient investment.

There is a need to rethink the educational processes in which teachers in this category can interact in a more human and affective way, because this is a different audience; when students see that the teacher is committed, they immediately become stronger and seek the strength to take part in the process of change. It is necessary for the educational community, through the Pedagogical Political Project (PPP), to be able to contribute to the evaluation process, making the process less archaic, in view of the fact that evaluation is about inclusion with ways of welcoming in a healthy, pleasant way. When this doesn't happen, it is characterized as exclusion, with no constructive power, thus leaving the student on the margins of society.

We need to make the students of Youth and Adult Education discover an enchantment for Mathematics, through creativity, contextualizing it with the reality of their daily lives. In this way, they will find hope in this school proposal, which aims to develop human talents.

This work is composed and organized into 10 (ten) chapters. In the introduction, we present a proposal for exploratory investigative practice, involving rational numbers, which can build a context favorable to the production of meanings for these numbers in the EJA.

The general objective is to analyze how students' knowledge of rational numbers is presented in the initial years of the EJA at school and in practical life.

Specific objectives include understanding the ideas, relationships and representations of rational numbers and the formation of concepts. This will lead students to locate themselves spatially; to identify changes in the mathematical registers of rational numbers and to use the registers of representation simultaneously.

In the second chapter, we explain how this study can minimize difficulties in understanding rational numbers. The study seeks to contextualize through practical exercises with the reality of the subjects involved in the research, bringing the fractions represented as close as possible to the reality of the community where the individuals come from.

On the national scene, the second chapter seeks to demonstrate the concern about what is expected of the individual when reading. According to Freire (1989), individuals not only need to learn to read and write, they also need to understand what they read and write what they understand.

The second chapter focuses on the experience in the classroom, with the challenges being much more challenging, because this time there is a deeper analysis of the initial difficulties in the class in the south zone, where there was even a need to change the methodology of teaching practice, slowing down the actions in the classroom, because there was a student with a hearing impairment, so I tried to interact directly with the student's interpreter[1] , investigating the level of understanding she achieved. I also investigated her background and the difficulties the interpreter was or wasn't facing in that situation. I soon realized that the information given to the student was very affective. Despite her friendliness, she had faced many difficulties in previous years; she also felt out of place in relation to the other students in the class, despite the fact that the audience also belonged to a special category, the EJA. However, there was a productive development in the conduct of the activities.

The second chapter also comments on the cognitive value of the diversity of representational registers, making it possible to change registers and treatment, since semiotic representations (Duval, 2009) enable certain cognitive functions to be fulfilled.

Finally, the second chapter looks at the importance of ethics in the development of educational activities, demonstrating that ethical pedagogical attitudes tend to produce more efficient and effective results, while authoritarian attitudes alienate and inhibit subjects, making them mere class attendees, where, however, classes can and should favor an exchange of information, always contextualizing with the daily lives of individuals, leaving them capable of producing knowledge.

The third chapter highlights the literature reviews, where the proposal is based mainly on Duval's (2004) hypothesis regarding the learning of mathematics, centered on the diversity of representations and the cognitive cost of these activities, which can denote two or three registers of representation of the rational number. In relation to natural, symbolic and figural language. The fourth chapter deals with the theoretical foundation on which the work is based, seeking in the investigation to develop alternatives to make knowledge of fractions meaningful, using the registers of semiotic representations. The methodology applied is commented on in the fifth chapter, which aims to qualitatively explore the association between the numerical representations

---

1 Interpreter - Responsible for the understanding of the math class of the student in the south class, as she has special hearing needs and needs help to better understand the teaching-learning process.

of rationals and decimals, in relation to the two classes, one in the northern region and the other in the southern region. The sixth chapter then discusses the search for alternatives to minimize or find alternatives to detect the problem of learning rational numbers. Using mechanisms and tools that produce significant results. The seventh chapter presents a data analysis using graphs with the first data from the research, following on from the results of the tests in the previous chapter. The eighth chapter shows the characteristics of the data collected in the previous tests, i.e. what each test contains. The ninth chapter deals with the analysis of the correct answers to the questions by class, presenting the results obtained in terms of the difference between errors and correct answers, in relation to gender. And finally, the considerations contained in chapter ten, where we comment on the difficulties and limitations of the research, the resources used, what we believe the school can contribute to achieving better performance in the teaching-learning process, not only in relation to rational numbers, but in a broader perspective. Finally, we hope that this study will not come to an end now, but that it will provide elements that will help to make the understanding of rational numbers more efficient, as well as in other areas of knowledge.

2   Understanding rational numbers in primary education

It is a fact that understanding fractions is a long-standing difficulty that students have been carrying around for some time as a heavy burden. With this in mind, this work tries to minimize the difficulties on this subject. EJA students have a greater degree of difficulty, given their time away from the classroom. Rational numbers were a complex and widespread subject even before Christ; they already provoked controversy and at the same time contributed to representations, because through symbols, pieces of bone were represented indicating broken quantities. When the River Nile flooded, the Pharaoh would divide up the land that was left over and distribute it to a certain privileged category, but nowadays there are various ways of giving students access to fractions with greater understanding.

After applying the tests with pictures, the student feels more comfortable doing the calculations, and says clearly without beating about the bush: "Now, teacher! I've understood. Written language, together with symbolic language and natural language, without avoiding grammatical rules or the expressions of symbols and figures, provide greater possibilities for understanding the object being studied. This is one of the important points for producing better quality in the aspect of fractions during the teaching and learning of mathematics.

This study seeks to contextualize the reality of the subjects involved in the research through practical exercises, bringing the fractions represented as close as possible to the reality of the community where the individuals come from. However, it is worth pointing out that however simple it may seem to represent the symbols, great care is needed in this task, first using natural language as the object of representation, then demonstrating the same object represented initially through a symbol, and finally through a figure, using figural language, where the object cannot be confused with its representations; in other words, the object must be recognized in each of its representations. It is hoped that the introduction of figural language will always provide a greater understanding through the use of calculations, because natural language, for an audience that has been absent from mathematics activities for a long time, the representation of the figure makes the connection between the other two registers of representation more efficiently.

According to Duval (2009), an activity specifically centered on the change and coordination of different registers of representation produces spectacular effects on the macro-tasks of production and comprehension. Mathematical activities allow for greater comprehension when the way the content is presented is increased, leaving the student with an extra benefit, i.e., in addition to a more attractive form, it becomes more accessible to the student. This promotes more efficient results.

2.1 A BRVE HISTORY OF YOUTH AND ADULT EDUCATION IN BRAZIL

Youth and adult education in our country was born out of the idea of reducing the number of illiterate people so that they could exercise their right to vote and become workers capable of producing more for the country. However, from Paulo Freire's perspective (1996), citizens not only need to learn to read and write, they also need to understand what they read and write what they understand. But unfortunately, even today in Brazil, youth and adult education functions as a depository for people who can acquire knowledge in any way.

teachers are not properly prepared to work specifically with these subjects, and students with some kind of disability are often included in the EJA, where the professionals have not received the training they need to conduct the teaching-learning process efficiently in parallel to the regular teaching of the class, resulting in unimpressive results, but which do not affect the school's iDEB. In this way, they continue to conduct the process, even using professionals who have not been able to solve the problems of young people in secondary school in order to deal with the EJA, which also requires differentiated treatment.

Although debates have been going on since the 1940s in the search for alternatives to develop specific education for adults, even today people involved in school education are still making speeches that cause embarrassment about this type of methodology in our country. It is with efficient and effective public policy measures that we can possibly turn this page that is so colorless, so expressionless, and that we can give a real right to this public that so desperately needs and deserves a minimum of dignity.

(MEC, 1997), emphasizes evaluation, final and certification criteria. While affirming the need to recognize the cultural diversity of young people and adults in literacy practices, the document with the curricular proposal for primary education covering this type of teaching, presents a selection of final assessment criteria for certification and future inclusion of students in the regular education system. These criteria are, in general terms: (understanding a read text, producing a written message, reading and writing natural numbers, performing calculations, solving simple problems and identifying information contained in simple tables or diagrams). So how can flexible proposals for valuing cultural diversity and the need for an assessment of cultural diversity and the need for a final assessment be reconciled with the competences indicated?

Once again we come back to the reflections of Café (1996, cited by BRASIL, 1999, p. 103), who illustrated the link between diagnostic assessment and the final assessment required for certification, in her work with the education of street children in Goiânia. According to the author, the diagnostic assessment was the focus of the experiment, with the final tests only being applied when the teacher was sure that the student could achieve more than 70%. It is important to note that the self-esteem of these young people was a constant concern for the team. So, in order to avoid "humiliating" those who were not yet ready for the final assessment, we tried to deal with the issue in the group, showing everyone "that everyone has their own pace of learning due to the different obstacles they have to overcome".

This and other similar experiences seem to indicate that, although we must question the existence of any centralizing proposal with the requirement of a final assessment and the indication of pre-established criteria for it, it is still possible to build spaces for valuing cultural diversity, for which the vision of assessment from a diagnostic perspective represents a relevant contribution. Otherwise, a final assessment will simply represent a classification that will condemn groups from different cultural backgrounds to failure and exclusion for the second time.

Assessment must be a continuous process that is always in motion. It is necessary to assess students every day, so that they are stimulated and their self-esteem is raised, as doubts are resolved during the course of the activities. In this way, the results are much more significant.

2.2 A CURRENT OVERVIEW OF THE EXPERIENCE WITH YOUTH AND ADULT EDUCATION IN PE.

At the beginning of the research, I told the students that during the course of the lessons, I would make some interventions to support the topic being researched and that, when the time came to introduce fractional numbers, I would like their collaboration, as I needed to find out more about the level of knowledge they had acquired in previous years.

At the beginning of the year, I had already talked about how those students were doing, and I realized from the start that the picture wasn't the best, because through surveys and small revisions, I identified a lack of ability with basic operations, which already signaled quite a challenge. In the class of EJA graduates, there was a special student who was hearing impaired. That's why I tried to slow down the speed of the information and interacted directly with her interpreter, investigating her level of understanding. I investigated her background, the difficulties the interpreter was facing or not facing that situation. I soon realized that the information given to the student was very affective. Despite her friendliness, she had faced many difficulties in previous years; she also felt out of place in relation to the other students in the class, despite the fact that the audience also belonged to a special category, the EJA.

It was noticed that, when they are well received, these individuals increase their self-esteem and their belief in a fairer world, with more love and solidarity. With cultural insertion, they can even regain interest in pursuing their dreams, which have been interrupted over the years.

I was now faced with yet another challenge, but to my surprise, the presence of this student made me see a new way of looking at things, which would only bring rewards for the others due to the methodology I would have to apply in order to obtain a greater return on learning. Little by little, I discovered that this type of audience needs a bit of affection to achieve more significant results, given the time away from school and the fact that the age group was already quite advanced, although she was carrying with her the hope of changing her life.

There is a double responsibility in conducting these classes, because the other classes don't evaluate the public with this sensitivity. In the development of the content, there is always a glaring deficiency in relation to conventional classes, thus showing that the teacher has to be prepared to face challenges of this nature. Despite the humility of the subjects in question, it was clear that when it came to claiming their rights, they were very determined about their objectives. It was quite clear that their main objective was to change their lives by entering the job market after completing their studies

With regard to the topic addressed to the EJA, I always tried to make it clear in all of them what objectives I wanted to achieve, because every day the question that kept coming up was the following: Where are we going to use this teacher? Will it be necessary in our daily lives? But it's also a reason for the proposed activities to always be geared towards producing efficient knowledge that can easily interact with the subjects' world.

The first evaluations didn't show significant results, but without wasting any time, quick reviews were applied, showing small flaws during the resolution of the activities, but what was observed in the failures was a lack of attention and concentration. I immediately tried to reassure them, but at the same time to warn them of the objectives to be achieved, because they require dedication and commitment, and when we don't act in this way, we certainly won't be able to achieve what was planned.

This small approach refers to the southern class, while the northern class has a school with greater organizational power, in addition to the physical space, and the political pedagogical project is more determined to achieve what is planned. In the EJA, however, the public is much more interested than in the southern class, as well as having a greater power of concentration. The teacher responsible for this class received the first tests with a lot of enthusiasm and promised that in the introduction to the research topic, fractional numbers would be covered very clearly. The students show a lot of interest, but their expectations of the objectives at the end of the course are no different from those of the class in the south. The subjects investigated here are striving to acquire an opportunity to enter the job market, increasing the possibility of helping their families and not being so marginalized by the programs offered by the job market.

The programs aimed at young people and adults not only don't offer teachers specially trained to work with this audience, but they are also not geared towards professional qualification. When this happens, the chances of a more dignified life on the job market will certainly increase. EJA students need daily motivation and commitment from teachers at all costs, because they play a different role from traditional students, who often don't value time as a fundamental factor in better learning.

It is essential to express our concern about this EJA group. The difficulties that each one brings with them are notorious. In small conversations, students gradually expose their difficulties, not only with mathematics, but specifically with rational numbers. We need our own way of dealing with this audience, not least because it is their right to be in a less unequal position than other members of society, since these individuals can also be capable of producing and competing with the various components of their community, allowing them to become a more dignified citizen and entitled to the respect of others.

These citizens to whom I am referring in the two northern and southern classes are people who feel excluded, who have high repetition rates, who have had their studies interrupted due to family problems and who work, most of the time providing services in their own community, but who in no way have a qualified education. These groups of citizens make it clear that their main objective is to obtain certification, in order to try to change their lives with an almost supernatural effort. It is necessary to reflect on the best way to serve such a special public, who carry with them marks that need to be mitigated.

After the rationale content was implemented in the classroom in both the north and south modalities, most of the students became more curious, as they created a certain expectation in relation to the results of the research. It was clear that in this type of teaching, the latent differences are hyper and exposed, as both the student and the teacher behave very differently at all times. Naturally, there is a greater relationship in terms of affection, given that the subjects surveyed show such sincerity in the course of the activities; that the teacher also changes his methodology, participating more closely in the intimacy of these students.

Based on this small approach, we believe that this public deserves more attention, as they feel as if they were excluded from a society that should welcome, motivate and encourage them to develop their skills, making them productive, thinking beings who can walk with their heads held high without any shadow of doubt as to their guaranteed and unshakeable rights. However, it is to be hoped that there will be more significant levels of investment to demystify this practice, and definitively fulfill what is de facto and de jure, not denying quality education.

Article 208 of LDB 9.394/96 states that it is compulsory to offer education to all citizens of the regular age group, as well as to those who are over the conventional age. However, it is difficult to include adults, as they are usually informed by relatives, friends and neighbors in their community. The media doesn't have enough mechanisms to attract this type of audience to attend school, and we need to develop effective and efficient public policies to get this audience to have access to school, sensitizing family members and the community in general, expanding the value and, above all, the importance of attending this type of education.

From the perspective of evaluation, it is necessary to go beyond simply measuring and assessing the apprehension of the content in question, which is simply transmitted in the search for the construction of knowledge. It is also necessary to assess the best way to integrate the student as a respected social being in his community, thus enabling him and his family to have an ideal quality of life. In this way, there will be social transformation and not merely the measurement of knowledge, because this knowledge in his community will take on new dimensions, because once this view is taken, the subject will have arguments that modify his behavior and that of the people who live with him.

2.3 A COGNITIVE PERSPECTIVE

In mathematical activity, according to Duval (2009), there is always the possibility, from a cognitive point of view, of checking for changes of register in semiotic representations, since there is a diversity of forms of representation of the same object. The originality of mathematical activity lies in the simultaneous mobilization of at least two registers of representation, or in the possibility of changing the register of representation, in this way the individual will require greater or lesser effort in certain situations to understand certain calculations, so when the object represented is clearly perceived, there is a cognitive evolution, enabling development without causing conflicts.

It is essential to have at least two different registers of representation in order not to confuse an object with its representation. Mathematical objects cannot be confused or misinterpreted in any way, as this can cause irreversible damage to understanding.

The symbols are arranged in such a way as to reduce the effort involved in understanding; in other words, the semiotic representations in activities involving mathematics in particular, when well represented, provide excellent results when calculations are required. Normally there are many difficulties when it comes to learning mathematics, but in this study natural language in the form of its representation will provide the subject with an economy of treatment in understanding that will help in the cognitive process. The aim is also to find ways to facilitate the understanding of rational numbers by contextualizing practical exercises with the reality of the subjects involved in the research, bringing the fractions represented as close as possible to the reality of the community where the individuals come from.

The forms of representation are fundamental sources for understanding when they are well represented, demonstrating significant gains in learning, the distinction between the object and its representation is a fundamental point for understanding it. In relation to understanding rational numbers, we tried to contextualize them through practical exercises with the reality of the subjects involved in the research, bringing the fractions represented as close as possible to the reality of the community where the individuals come from, making them more comfortable in producing results.

2.4 PROFESSIONAL ETHICS IN THE CLASSROOM

Ethical principles are essential for good coexistence, so the results will emerge, otherwise evasion will set in, discouraging the subjects, the public needs values to be rescued. Let there be a dialogical relationship, promoting the construction of knowledge, where there is never a transfer of data to the classroom blackboard; or a list of exercises, without first leaving a good basis of content with a lot of clarity and objectivity that encourages them to develop the exercises, early on you can see in the EJA a seriousness and a greater desire to acquire some knowledge.

It is essential, especially in a classroom of young people and adults, to apply ethics with a greater degree of sensitivity, along with a dose of affection. The subjects who take part in Youth and Adult Education (EJA) arrive at school exhausted, because most of them are low-income workers. This is why we have to consider the principles of attendance, organization, planning of previous activities, didactic material, with support and of a good level according to their preliminary difficulties. Trying to create a friendly and welcoming environment so that the fruits of this relationship can be reaped in the teaching and learning of mathematics.

Ethical pedagogical attitudes tend to produce more efficient and effective results, while authoritarian attitudes alienate and inhibit subjects, making them mere class attendees, where, however, classes can and should favor an exchange of information, always contextualizing with the individuals' daily lives, always leaving them stimulated to become more productive and giving meaning, where they should apply those teachings that are being produced in class. These attitudes only benefit the subjects and bring them closer to their dreams. One cannot use an authoritarian nature, where the teacher is unable to introduce freedom of expression as a form of teaching and learning.

The human side speaks louder, this citizen can't be treated as an underdog, but as a subject who not only has the right to be in that space, but also needs to be welcomed in order to change the current reality of the class they live in. Ethical principles lead students to rescue and apply them in the community they come from. The prospect of change is latent, they are not going to school by chance, the objectives are bolder than when most young people go to school, their time is different, it has become a luxury item. Ethical attitudes are therefore key factors in enhancing teaching and learning, especially in the EJA. Ethics is linked to teaching practice, without it there is no discipline, no planning, no harmony, no team, in short, ethics is an essential factor in achieving significant results in education. The ethics of the teacher in the classroom is similar to what Freire (1996, p. 37) says: "Ethics becomes inevitable and its possible transgression is a disvalue, never a virtue. It is not possible to think of human beings as far from ethics, let alone outside of it".

The school effectively participates in the construction of knowledge for life and the formation of the human subject and citizen. Individuals already bring with them ideas of limits and values, but it is necessary to make it clear to them that ethics is not a set of pre-established rules. But the student should be able to build, together with the teacher, a pleasant, stimulating, captivating space in the classroom, where subjects can experience skills and abilities, but above all, where they can be inserted into this space, where they themselves have taken part in the construction process.

Educating today requires, above all, an inclusive vision, seeking to understand the sensitivity of the subjects, thus valuing ethics, promoting that our students build solidarity, respect for cultural diversity. We must try to adapt the subjects to the place where they live, their community is a point of reference for discussing ethical and moral processes, because that is where they are, so there is already a behavioral form, which needs adaptation, thus seeking to make the environment as pleasant as possible, stimulating an ethical culture.

3 BIBLIOGRAPHICAL REVIEWS

The activity proposed to the students is based on Duval's (2004) hypothesis regarding the learning of mathematics, centered on the diversity of representations and the cognitive cost of these activities, and which can denote two of the three types of representations of rational numbers: the symbolic-numerical (fractional and decimal) or algebraic register; the figural register (representation of parts of discrete or continuous quantities); and the natural language register.

At the beginning of my post-graduate course, I had the opportunity to meet Professor Ademir Ferraz; where he spoke about Learning Theory. At the time, he cited various theorists, but the one that most caught my attention was Duval (2004); the main reference in the Bibliographical Review. It is important to think that the teaching of mathematics in primary, secondary and EJA schools does not aim to form great mathematicians, nor does it aim to store knowledge that can only be used in competitions or in the future, in higher education.

Above all, it is necessary to contribute to their development in a broad way that can encompass their reasoning and logical abilities. Hence the need to enable cognitive development so that they can think in different ways, understand and produce concepts that can be useful in everyday life, whether in the simplest of tasks, such as measuring the area of a room in your house to lay a new floor, or helping your family with an investment calculation that requires discounts involving percentages.

Because numbers are capable of exposing the problematization of countless situations. For example, building tables and graphs providing data on the current situation in our country, in relation to violence in schools, prejudice, the appreciation of women in the job market through their inclusion in construction; and also helping with basic household chores, guiding family members how to obtain significant discounts in negotiations, thus even generating income, reorganizing their own family, because numbers have the power to even size spaces in a more appropriate and economical way, even providing well-being.

According to Duval (2009), the analysis of cognitive development and the difficulties encountered in learning are confronted with three interconnected phenomena concerning the learning of rational numbers.

1-Existence of different semiotic representation registers - When rational numbers are introduced in elementary school, they are represented by the three types of representation register mentioned by Duval (2004): the symbolic register - numeric (fractional and decimal) or algebraic; the figurative register (representation of parts of discrete or continuous quantities) and, of course, the natural language register. Examples of these registers are shown in Table 1.

> The production of an answer, that of a text or that of a diagram, simultaneously mobilizes the formation of semiotic representations and their processing. The comprehension of a question, a text or an image mobilizes the three cognitive activities of conversation and formation (DUVAL, 2009, p.54).

These are not simply the activities of transforming representations within the same register, i.e. treatment. Neither is changing registers while maintaining the same objects, conversion. However, these are more complex resources within semiotics that require a study of the rules of conformity. Transiting between the different registers of representation is no simple task, because each register of representation has its own specific characteristics and if the student is unable to transit between the different registers of representation, it will be difficult to overcome the difficulties.

You won't understand what you're doing, much less be able to conceptualize the object you're studying.

Figure 1 - Register representation and rational numbers

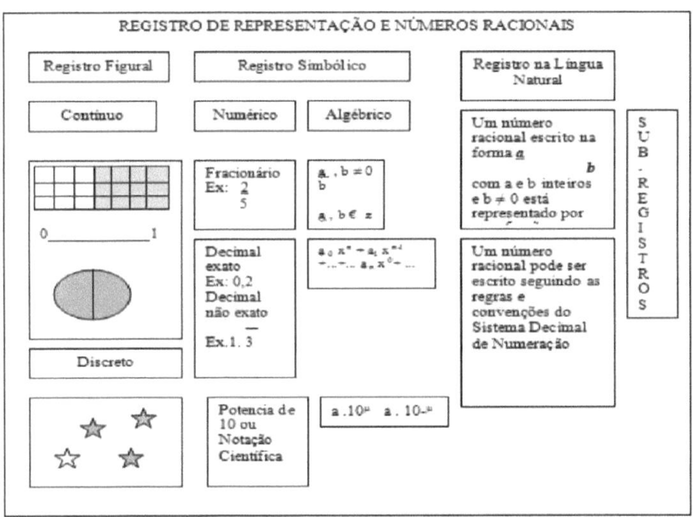

Source: MACHADO, (2009).

However, although the records are presented in the three forms of representation, they are not always well assimilated by the students, most of the time generating conflicts. Various forms of processing are needed to materialize this information, and then the results appear.

2- Differentiations between the represented object and its semiotic representation registers - Difficulties in learning related to the phenomenon of differentiation between represented and representatives can be detected in the following examples, according to Catto (2000). In the activity in which the task was to place the $=$ or $/=$ signs between $(0.5)^2$ and $(1/2)^2$, Elaine, a secondary school student, gave the following answer: (0.5) 2 = 0.25 $/=$ 1/4 =0.25 (Elaine's solution).

The fact that the student didn't identify the two representations as being of the same number is an indication of the occurrence of this phenomenon. In the chapter in which Catto (2000, p.71) analyzes textbooks, the following proposition is stated in one of these books: "When one or more zeros are added to or removed from the right of the decimal part of a decimal number, that number does not change".

We note that the word number (natural language register) is used interchangeably for both the object and its representative. From this register perspective, the phrase when you add (...) a decimal number, that number doesn't change (CATTO, 2000) can be understood as: when you change (...) a decimal number, that number doesn't change. A treatment in the natural language register - which would make the student realize that one thing is the object (number) and another thing is its representative (decimal form) - would be necessary, replacing, where appropriate, the word number with representation.

Most of the time, students confuse the mathematical object to be studied with its representation, which is very common. When students use the different registers, they don't confuse the content of a representation with the object represented. Hence the importance of the student having at least two registers of representation of the same object. 3- Coordination between different registers of semiotic representation - An activity that requires the student to write 0.25 as ¼ involves a conversion. In this case, the starting register is numeric, in decimal form, and the ending register is in fractional form. Another that asks, for example, for the equivalence between ¼ and 2/8 (numerical register in fractional form), is considered a treatment.

It's important to note that knowledge of the rules of correspondence between two registers may not be enough to mobilize and use them simultaneously. A student may know, for example, that they have to divide 1 by 4 to get the decimal representation of the rational ¼, but they may not recognize 0.25 as representing the same rational number.

As presented by Duval (2009), there is an obstacle to spontaneous coordination between registers related to the phenomena of non-congruence between representations of two semiotic systems. In Catto's research (2000), it was found that, in general, in elementary school, conversions are used less than treatments and, when they are used, one of the senses is prioritized.

Conversions are the most effective register changes for acquiring a concept. In particular, there is a greater chance of mobilizing students' knowledge in order to acquire the concept when conversions are made in both directions and are not congruent (they are not natural, direct).

But the difficulty is not necessarily the same in both directions of a conversion, as illustrated by the answer given by student Elaine: 0.25=1/4 and 1/4= 0.25. In one sense, when she saw 1/4 = 0.25, she identified the two representations as the same number. We credit this to the use of a known rule for this conversion (divide 1 by 4). In the answer 0.25 = 1/4, it seems that the student didn't have a rule at her disposal for converting 0.25 to 25/100 (congruent) which would only require treatment (simplification), in this register, to arrive at 1/4.

In this same task, Rafael (a high school student) spontaneously used the conversion feature from the numerical register (1/4) to the figural register (circle divided into 4 equal parts, with one of them highlighted). It could be interpreted that he considered the meaning of 1/4 to be that of part-whole, which ensured congruence between the fractional numerical register 1/4 and the figural one he chose. However, there was no congruence in the conversion of the decimal register as the same number.

The student spontaneously resorted to another register, which is desirable, according to Duval (2009), but not appropriate. The lack of congruence in the conversion from decimal to fractional representation seems to have led Rafael to consider 0.25 = 1/4. It is necessary to understand the conversion in both directions. This makes it easier to understand.

Sometimes one register initially proves to be enough to demonstrate knowledge of a particular object, but it is important, according to Duval (2004**)** **to** move from one register to another, because in order to really establish understanding in mathematics, it is necessary to coordinate at least two registers of semiotic representations.

According to Duval (2004, p.17) "The conversion of representations, whatever the registers considered, is irreducible to a treatment (in other words, if transformation by conversion occurs, there must be a change of register). "

According to Duval (2009, p.59) ". With regard to numerical calculation, even at the highest levels, students fail not in the processing activity, but in the conversion activity. Even if they know how to add two numbers in decimal and fractional form, some students don't bother at all to think about converting the decimal form of a number into its fractional form (and vice versa), or even fail when they realize that this is necessary in the development of a calculation.

In reality, decimal writing and writing with display constitute three different registers for representing numbers. In fact, when writing a number, it is necessary to distinguish between the **operative meaning attached to the signifier and the number represented.** So the operative meaning is not the same for 0.25, for ¼, and for 25.10^{-2} . Because it's not the same processing procedures that make it possible to perform the following three additions:

0,25 + 0,25 = 0

1/4 + 1/4 = ½

$25 \times 10^{-2} + 25 \times 10^{-2} = 50 \times 10^{-2}$

Each of these three signifiers "0.25", "1/ 4" and "25×10^{-2} " has a different operative meaning, but represents the same number. If the operative meaning attached to the signifier and which commands the treatment procedure is not differentiated from the object number represented, then the substitution by convention of 0.25 for 1/4 is no longer conceivable!

In this way, there are no set rules stating that a convention activity requires the same degree of difficulty as a treatment activity. For a given student can convert the understanding of a graph, using figurative language to a symbol with a certain ease and vice versa; but another subject may experience a greater degree of difficulty in this requirement, the difficulties may be ambiguous, but the order will not characterize a greater or lesser cognitive development in terms of scoring.

However, these activities, whether or not they require certain efforts to make sense of the changes in the treatment of these registers, should not interfere with the conceptual understanding of the object studied. According to Catto (200, p.54), a great contribution is made through a proposed exercise, where the figure illustrates fractions and provokes the visualization of the figure, where it allows the condition of a conversion, it is necessary to carry out a treatment in the figural representation in order to simultaneously recognize the equivalence between the portion corresponding to the fraction ¼ and the two fractions 1/8.

The addition operation in the fractional register involves converting the figural register into the fractional register. Therefore, in this exercise, there are two activities involved, the conversion and the treatment; in this way, two treatments are carried out, one in the figural register and the other in the fractional register, which makes it possible to obtain the result of the operation in the fractional register.

Table 1- Exercise with fractions

$\frac{1}{8}$	$\frac{1}{4}$
$\frac{1}{8}$	
	$\frac{1}{4}$

Source: CATTO, (2000)

In the figure, the author proposes the addition of fractions, with the aim of observing the change of register of representation, in the possibility of conversation; when she asks the student to pay attention and answer the following additions:

a) $\frac{1}{4}+\frac{1}{8}$

b) $\frac{1}{4}+\frac{3}{8}$

c) $\frac{1}{8}+\frac{2}{4}$

d) $\frac{2}{4}+\frac{4}{8}$

In this situation, the student tends to realize that there is a relationship between the representation of the fraction of ¼ in equivalence with the two of 1/8, because the figure provokes this understanding, and it is important to note that there may be more or less effort in relation to this understanding. It is precisely this transition that helps students to advance cognitively in their understanding of fractions. The links between these registers are essential for a well-defined understanding of a rational number.

4   THEORETICAL BACKGROUND

It is clear that teaching has been the subject of constant study and research aimed at developing theoretical support for understanding the teaching-learning process in mathematics. In this case, the assessment tools cannot be just any old tools, but ones that are suited to what is intended to be achieved at the end of the activities developed, establishing clarity so that the students understand what they can provide as an element that allows them to visualize the results naturally. The challenge of the research was to find an alternative to make knowledge of fractions meaningful.

It is therefore essential to create concrete tools and make them available, so that ideas can happen and be understood at the time.

> The originality of mathematical activity lies in the simultaneous mobilization of at least two registers of representation at the same time, or in the possibility of changing registers of representation at any time (Duval, 2004, p.14).

In Mathematics, the effectiveness of the semiotic representation registers is more evident, as it makes clear the evolution of thought in this discipline. Duval (2004) seeks to clarify the importance of the need to understand the semiotic registers that subjects reveal in learning, as they are observed in the student's cognitive development. Duval (2003) classifies semiotic representation registers into four distinct types: natural language, geometric figures, writing systems and Cartesian graphs, as shown in the table below.

Chart 1 - Classification of the different registers that can be mobilized in mathematical functioning (doing mathematics, mathematical activity).

	Discursive Representation	Non-discursive representation
REGISTRATION MULTIFUNCTIONAL: The treatments are not algorithmizable.	**Natural language** • Verbal (conceptual) associations. • Formaderaciocinar: • argumentation based on observations, beliefs...; • valid deduction from a definition or definitions or theorems.	**Flat or perspective geometric figures (1, 2 or 3 configurations)** **D)** • operative apprehension and not just perceptual; • building with instruments.
MONOFUNCTIONAL REGISTRATION The treatments are mainly algorithms.	**Writing systems** • numeric (binary, decimal, fractional, ...); • algebraic;	**Cartesian Graphs** • changing coordinate systems; • interpolation and extrapolation

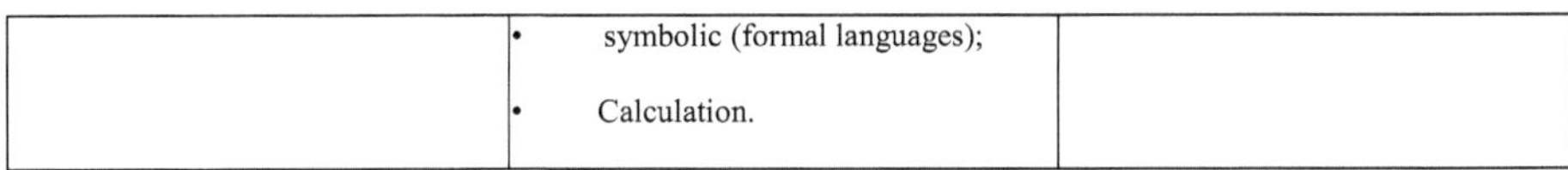

	symbolic (formal languages); Calculation.	

Source: DURVAL (2003, p.14)

In this way, discursive representations exert power over the reading of language in order to demonstrate, interpret and create a more global vision, while non-discursive representations immediately make it possible to know how to do things, through visualizations of instrumental power.

The two registers promote a transformation, i.e. treatment and convention, causing a transition between one register and the other, as can be seen in the table below.

Table 2- Two types of transformation of semiotic representation.

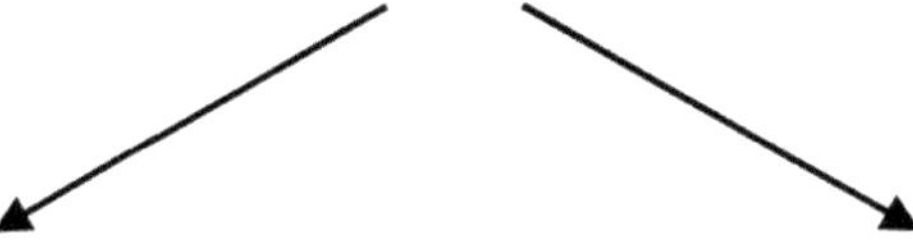

Transformation of one semiotic representation into another

Remaining in the same system: treatment.	Changing the system, but keeping the reference to the same object: conversion.
Almost always, it's only this type of transformation that attracts attention because it corresponds to justification procedures. From a "pedagogical" point of view, we sometimes try to find the best register of representation to use so that students can understand.	This type of transformation faces the phenomenon of non-congruence. This means that students don't recognize the same object through two different representations. The ability to convert implies the coordination of mobilized registers. The factors of non-congruence change according to the types of register between which the conversion is, or must be, carried out.

Source: DURVAL (2003, p.15)

It is important to emphasize that the tables above show that the four semiotic representation registers and the transformations between them make it possible to advance learning in EJAI math classes, distinguishing the object in relation to natural language and symbolic language.

According to Duval (2003), the transit between these two registers takes place when the specific content is identified and coordinated with the object being studied. And he makes it clear that when the subjects surveyed perceive the figure, it is immediately contextualized with the first, thus materializing knowledge.

In this sense, it is important to develop activities that lead students to a good understanding, without causing conflict in their understanding. Representations are fundamental in the development of the construction of knowledge, given the diversity of interpretation of the representation. According to Duval (2004), an adequate understanding of the representation is necessary. That's why it's important to develop an attitude that favors the teaching and learning of mathematical content, it's a process and it needs our investment. Therefore, we need to review our attitude, our relationship with the classroom, the way we relate to our students' learning process is fundamental.

Therefore, it is important to master the content to be covered, studying it in advance, intervening and contributing to the organization of the students' records. The teaching of operations with fractions should use their various meanings and link the subject to the students' daily lives, arousing their interest and making the teaching-learning process more efficient and enjoyable.

It is very important that when developing a solution to a problem, there is not only a mechanical action to solve the problem, but that the student thinks when determining the solution. Thus, heuristically speaking, efficient productivity is a matter for the cognitive system, leaving students with the opportunity to produce their own concepts.

In relation to the representational system, there is content that can enable and enrich knowledge. Because when at least two pieces of information are recorded, we can obviously really understand an element and thus turn it into a tool. We mustn't forget that school is geared towards life, and must take into account the fact that students need to apply knowledge in their daily lives, whether they're measuring an area of their home, or making payments and needing to fractionate through discounts. In this way, they materialize the content covered in class. However, the students have great difficulty when faced with situations involving the use of operations with fractions.

Despite various studies, there is still very little impact in the classroom, with students still not understanding and having doubts about the two ways of representing an object. This makes it difficult for them to develop adequately and progressively in the subsequent grades, and therefore, when they reach a higher level of education, the bottleneck appears, thus hindering the development of skills that could take them a step further up the intellectual ladder.

In view of the above, we tried to work with the object of investigation, from an evaluative perspective, exploring the difficulties in understanding rational numbers, with the aim of supporting the quality of basic education in the early grades of elementary school. Raymund Duval's theory (2004) is an ideal theoretical reference point for the development of mathematics education. Above all, it is important to emphasize that when the student converges different types of semiotic representation, especially with regard to rational numbers, it opens up horizons for developing activities related to practical life; as well as measuring areas, having security when needing to divide, fractionate, in reading data with the family the option in the face of financing ready for the adversities of today.

The table below shows an analysis of the basic education development index for the state of Pernambuco, specifically for one of the schools where the study was carried out. It is worth pointing out that despite the results presented, the data offered by the research is not verified in practice. However, it is through activities such as this study that the current situation is changing.

But it depends on joint action, not just a few isolated cases. However, we are still developing tools to transform this reality, because it doesn't depend exclusively on public authorities, but also on educators.

Table 2 - EBs observed in 2005, 2007 and targets for the state network - PERNAMBUCO

Teaching phases	Observed IDEB		Projected targets							
	2005	2007	2007	2009	2011	2013	2015	2017	2019	2021
Early Years of Elementary School	3,1	**3,5**	3,2	3,5	3,9	4,2	4,5	4,8	5,1	5,4
Final Years of Elementary School	2,4	**2,5**	2,4	2,6	2,8	3,3	3,6	3,9	4,2	4,5
High School	2,7	**2,7**	2,7	2,8	3,0	3,2	3,6	4,0	4,3	4,5

Source: Saeb and School Census. JOSÉ GLICÉRIO SCHOOL

Despite various studies, the effects in the classroom are still very limited, with students still not understanding and having doubts about the two ways of representing an object. This makes it difficult for them to develop adequately and progressively in subsequent grades, and therefore, when they reach a higher level of education, the bottleneck appears, thus hindering the development of skills that could take them another step up the intellectual ladder.

In view of these circumstances, the object of this research is: difficulties in understanding rational numbers, with the aim of supporting the quality of primary education in the early grades of elementary school and the EJAI. Raymund Duval's theory (2004). It is ideal support as a theoretical reference of great value for the development of mathematics education. Above all, it is important to emphasize that when the student converges different types of semiotic representation on us, especially with regard to rational numbers, it opens up horizons for developing activities relating to practical life; as well as measuring areas, having security when needing to divide up, fractionating when reading data and family choice in the face of financing ready for the adversities of today.

5 **METHODOLOGY**

In order to qualitatively explore the association between the numerical representations of rationals and decimals, two EJAI high school classes were chosen, in which a questionnaire was first administered to 32 to 35 students, located in Jaboatao dos Guararapes and Paulista, Pernambuco, involved in the research, with the aim of identifying the intervening variables, such as: the level of knowledge about rationals whether they like mathematics, class components, how they entered school, age, gender and whether they are repeaters. (Appendix A).

The purpose of identifying these variables was to analyze their interference in the results of the tests applied to the students in the two classes involved in the research, before and after the tutoring classes on the pre-established days.

The first test applied, before tackling the content to be analyzed in the research, the association between the numerical representations of rationals and decimals, consists of five objective questions, as shown in Appendix C. These questions were prepared taking into account that this content is part of the syllabus for the 1st grade of secondary school. [a]

After the first test, the classes began. The first class, class 01, located in the town of Jaboatao, PE, began applying the content at the same time as class 02, located in the town of Paulista, PE. First of all, the fraction ruler was introduced, a tool used to break down expectations and anxiety about how the lesson would start. The tool helps us to understand the parts of a whole. From then on, we developed the content of the rational itself. During the lessons we tried to show symbols, figures and practical exercises. The assessment was geared towards enabling figural and symbolic language, with the aim of seeking a lower cognitive cost in carrying out the task, in order to broaden the field of vision.

According to Duval (2004), the existence, in this activity, of the figural register and the fractional register of the rational number suggests to the student the use of a change of register that provides a lower cognitive cost in carrying out the requested items. What's more, the use of the two registers makes it possible to work with two reference systems that have specific internal rules, broadening the scope for learning the mathematical object.

Representations are fundamental in the development of the construction of knowledge, given the diversity of interpretation of the representation. The students gradually demonstrate a gradual progressive development, which fuels stimuli to develop the learning process.

6 Speech

Attempts have been made to find alternatives to the problem of learning rational numbers. In the search for an alternative approach, but with the use of tools that are already common in our environment, and also among the majority of students; such as the fractional ruler, the fractional disk, the abacus, the tangram, the golden Chinese game, in short, there is a wide range of materials that must become necessary tools and are developed through creativity and that make it possible to do a good job. Most students have great difficulty learning fractions; they are often unable to recognize whether 1/3 is greater or less than 1/4. Fractions involve several ideas and all of them should be worked on well in the classroom.

Some students acquire incomplete notions and may even learn how to add or divide fractions, but in a mechanical way, without really understanding what they are doing. This is why they end up making mistakes, for example: 5/3 + 1/4 = 6/7. When faced with situations like these, the teacher must reinforce the first basic ideas that gave rise to fractions.

The following pictures show that when the class is introduced to equal-sized pieces of cardboard, the student is asked to fold the pieces in such a way as to divide them into 2 or 4 or 8 equal parts.

Figure 2- Fractional ruler

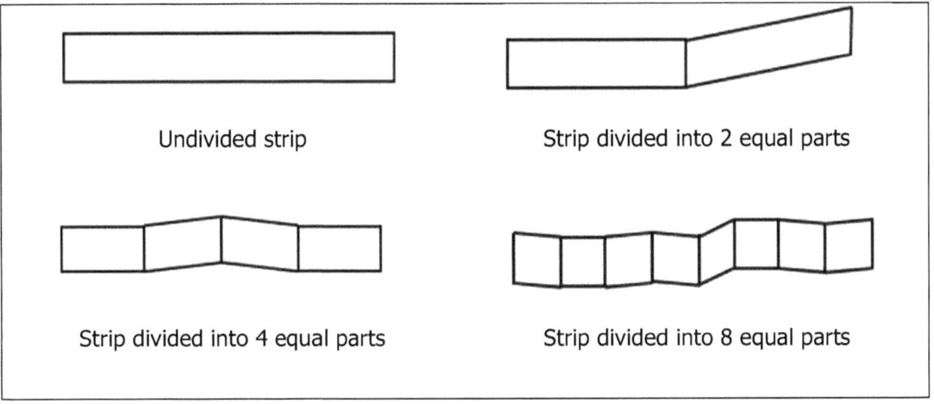

Source - the author

Once the pieces have been completed, i.e. the half is found, then the half of the half; the student is encouraged to try to find the others and so on. They are then shown the fractional ruler, which is made up of whole parts and countless pieces that gradually form the whole.

In this way we avoid memorizing definitions and rules without understanding them, and then we can visualize the representation of the object. Let's take a look at an example of the use of the fractional ruler and its importance during lessons:

Table 3 - Example of a fractional ruler

<table>
<tr><td colspan="16">1</td></tr>
<tr><td colspan="8">½</td><td colspan="8">1/2</td></tr>
<tr><td colspan="4">¼</td><td colspan="4">¼</td><td colspan="4">1/4</td><td colspan="4">1/4</td></tr>
<tr><td colspan="2">1/8</td><td colspan="2">1/8</td><td colspan="2">1/8</td><td colspan="2">1/8</td><td colspan="2">1/8</td><td colspan="2">1/8</td><td colspan="2">1/8</td><td colspan="2">1/8</td></tr>
<tr><td>1/16</td><td>1/16</td><td>1/16</td><td>1/16</td><td>1/16</td><td>1/16</td><td>1/16</td><td>1/16</td><td>1/16</td><td>1/16</td><td>1/16</td><td>1/16</td><td>1/16</td><td>1/16</td><td>1/16</td><td>1/16</td></tr>
</table>

Source - the author

After presenting the fractional ruler, the students are told to paint each part. With a different color, they cut out the parts of one of the sheets. They are asked to check how many halves are needed to get an integer; how many quarters are needed to get a half; how many sixteenths are needed to get an eighth. Secondly, they are asked to add ½ plus 2/4, for example, and present the result. To do this, use the sheet with the whole fraction ruler and the parts cut out. In this way, the students will probably come to the conclusion that ½ plus 2/4 is an integer and that we need 2/4 to get ½.

In this way, the student begins to understand the fractional numerical form and the symbolic register; a distinction is made between the role of conversation from a mathematical point of view and from a cognitive point of view. From the mathematical point of view, the conversation consists only of switching to a more economical register, and does not play an intrinsic role in the mathematical processes of justification or proof.

It is treated as a side activity, but from a cognitive point of view it is the conversation, because it plays a fundamental role. It leads to the underlying operations of comprehension, which we want so much that we advance every day, aiming for a student who has the power to produce concepts, create possibilities to solve problems, in short, to be able to produce in general.

There are some other concrete materials used as teaching aids to facilitate understanding in the teaching-learning process, but it is necessary to point out that there is already controversy over the use of these materials, such as some games. It's important to realize that the most important thing when using materials are the skills that the teacher, in possession of such a tool, uses with great skill. That's why it's necessary to follow up continuously, step by step. That's the only way to find results in the short term, which is what you want. Otherwise, misinterpretations arise and can cause endless conflicts, often even irreversible ones.

It is through the meaning conveyed that reflections are made on the actions developed by the teacher, which will possibly lead to advances in the construction of knowledge and the development of concepts. It can be seen that the contributions made using the fractional ruler were clearly satisfying, and this can be seen in the interaction between the other classmates. The class was then encouraged in general to comment on the progress made.

After this explanation of fraction strips and the ruler, we proposed some practical problems from the everyday reality of these subjects in order to understand the transformations through the registers of representation; now investigating whether a greater degree of understanding is really established when the student is offered another type of object to work with in the different classes.

In this way, the first challenge was proposed, initially without the use of figures and with few symbols, initially assessing how many they can develop without the aid of a representation. These tests were designed and presented to the EJAI students in two ways, the first of which does not contain much information, but has

The second problem has the same context, but contains data from figures, thus facilitating the transition between the registers of representation. The second problem has the same context, but contains data from figures, thus facilitating the transition between the registers of representation, seeking a better understanding of the object of study, according to (Appendix D).

This allows students to understand and interpret the information in the figures, so that they can possibly move through at least two semiotic representation registers, according to Duval (2003), i.e. when the student is able to identify a representation register for the mathematical object being studied.

So the first test has five (05) questions, as already mentioned, and investigates what percentage of subjects in the universe of 10 students in a class of 32; how many have the ability to develop and present results using at least two semiotic representation registers.

Graph 1 - % correct without figure class south

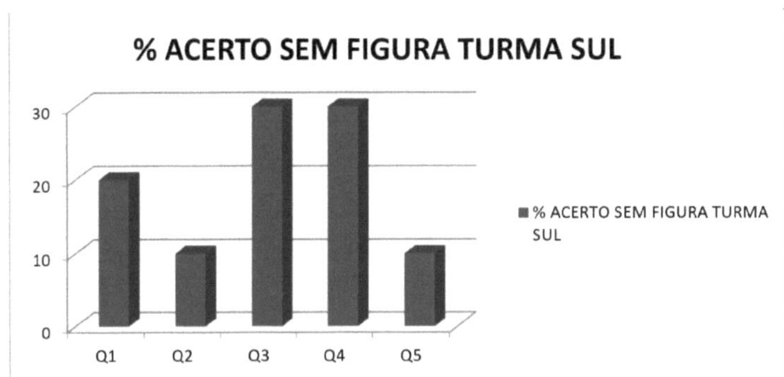

Source - the author

After the tests were administered, there was a lot of discussion and endless quizzing, with the majority asking as usual: teacher! Will it hurt me if I don't do well? The results were not very expressive, but they will be shown later in the data analysis. It was clear that the approach, where only the writing appears more rigorously, naturally shows a low level of comprehension, given the difficulty in interpreting the reading.

27

Firstly, it becomes difficult to understand if we don't distinguish the represented object, which Duval makes clear in his contributions to mathematics, because when represented they tend to dispel any misunderstandings, not leaving the subject confused.

The new proposal aims to represent the objects described above in greater detail, trying to establish a more realistic approach to the subjects' understanding. The problems are now illustrated, providing a greater level of comprehension, given that there are more registers of representation.

Comments and analysis of the test results of the second class in the northern zone.

1- 30% of the students managed to successfully solve the first problem without help, even scribbling some figures to arrive at the final result.

2- In this second exercise, 20% did it correctly, arriving at the result.

3- - 40% of the students achieved results that were reasonably much more significant than the previous ones, noting that in the problem request the understanding was better accepted.

4- In this problem, the 40% of students with positive results remained, as the exercise even maintained a certain similarity to the previous one, perhaps not requiring much effort to interpret.

5- Only 20% of the students managed to solve this exercise successfully, although the exercise requires similar attention to the first one, but it can be justified or not by a lack of attention.

Graph 2 - - % correct without a figure - North class

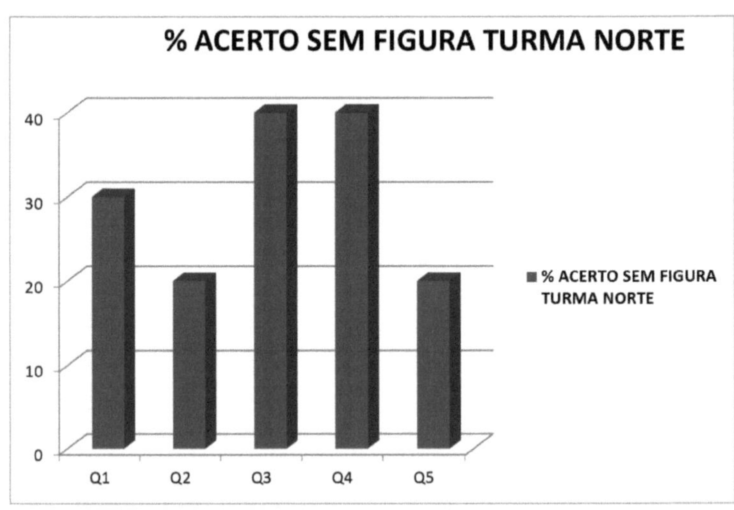

Source - the author

This second test has well-illustrated figures with a wealth of details in an attempt to make it easier to understand for the same subjects who took part in the first one, thus investigating whether in this second condition the results are more expressive and can also make the idea of the object studied clear; because fractions always appear as a stigma, when it comes to breaking, dividing, adding smaller parts, difficulties always appear.

In this other model of presenting the problems, the level of acceptance and understanding had a very positive impact right from the start, because the figure naturally brings greater certainty in the level of transformation in the representation of the object.

Comments on the application of the second test, class South:

1 - In this new format of preparing the same test, but with a greater amount of information and richness of detail, because another language is figural, the student doesn't need to mentally request the figure, because it is already there, facilitating reasoning, so 40% of the students solved the problem.

2 - In the second problem, even with the help of figural language, only 30% of the subjects managed to successfully solve the problem and arrive at the correct solution, as this problem is still presented initially with the fractional part, requiring greater attention from the student.

3 - In this third challenge, only 40% of the students managed to solve it successfully, because the presentation of the exercise helps to make solving it more practical, requiring reasoning, but with more simplicity.

4 - The fourth challenge was successfully solved by 40% of the students, as the level of difficulty was similar to the previous exercise.

5 - Only 30% of the subjects successfully solved this exercise, as it requires a little more attention, since the question starts with the fractional part, thus requiring a greater mental effort to reason.

Graph 3 - % correct with figure class south

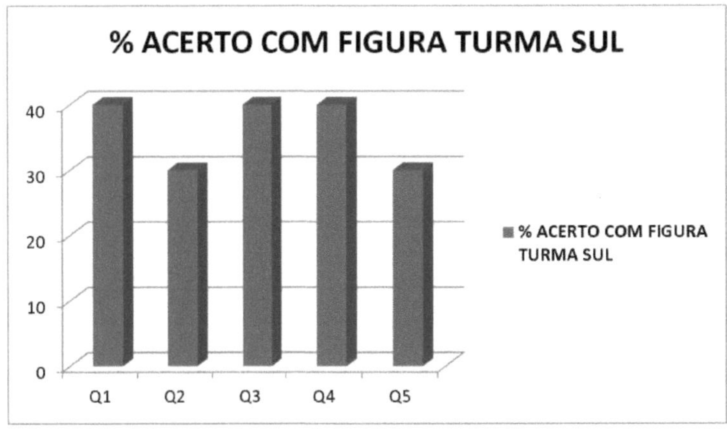

Comments and analysis of the second test in relation to the north zone class.

1 - In this first problem, 90% of the students were able to solve it successfully, reaching a satisfactory result, even using the figure very well, showing how the form of representation can facilitate the understanding of an object, taking the subject to a level of understanding with a greater degree of relevance.

2 - In the second exercise, 60% of the students achieved the objective, reaching a satisfactory result; however, it is clear that the way in which the exercise was presented required a greater degree of difficulty.

3 - In this third test, the students got 70% of the questions right, always pointing to the figures, which shows that they used the second register of representation as a facilitating factor in understanding the question.

4 - 80% of the students got the problems right, a very significant number, demonstrating a good level of attention.

5 - 60% of the students managed to get the exercises right, even though they were similar to the first exercise they were asked to do, as the exercise required a degree of difficulty very similar to the first one.

Graph 4 - % correct with figure - North class

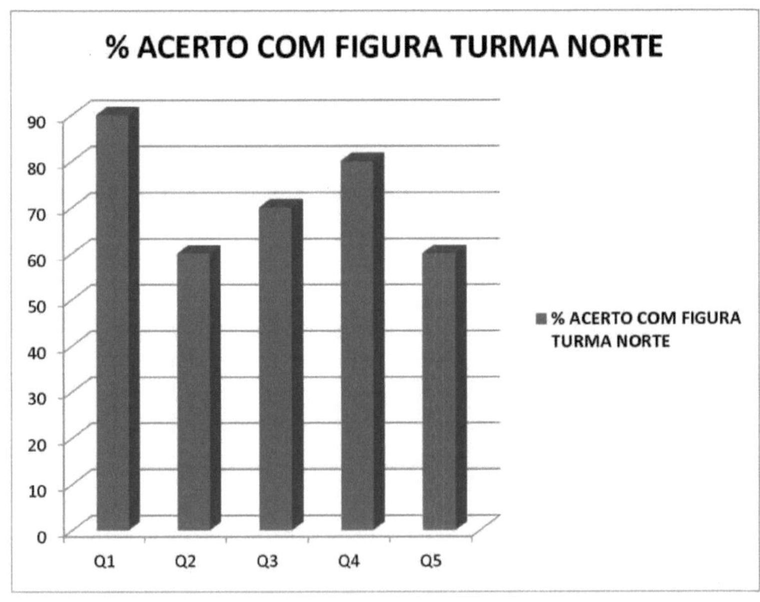

The results obtained in the second test were more representative in the formation of the problems, even with the help of the figures, providing details that certainly enabled greater understanding. These tests boosted the students' self-esteem, because previously the results had made the boys very worried, even using questions like, "Gee! I don't think I can learn this subject.

It was clear that the northern group received the most information, with the results even showing a greater concentration, where the results are much more significant than the first group.

In the indications of the graph analyzing the two classes, north and south respectively, it can be seen that the north class presents more expressive results than the south class; these analyses in more detailed conversation with the way in which rational numbers are conducted during the lessons, have some particularities.

The northern class is more assiduous than the southern class, as well as having a greater power of concentration during the development of the content, as well as when solving the problem, traces are found dividing figures in the search for a better interpretation of the proposed problem. It is worth noting that the northern class clearly demonstrated that they were always looking for the figure when solving the question, thus showing that there was a conversion of registers, so there was a transition to more than one register of representation. That's not to say that the southern class didn't show this, because there was also a change, but it was more modest.

Graph 5 - % error class north vs class south without figure.

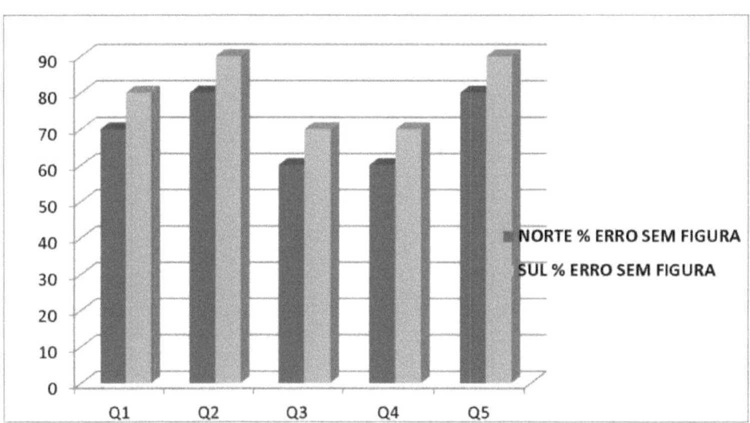

Source - the author

Graph 6 - % error class north vs class south with figure.

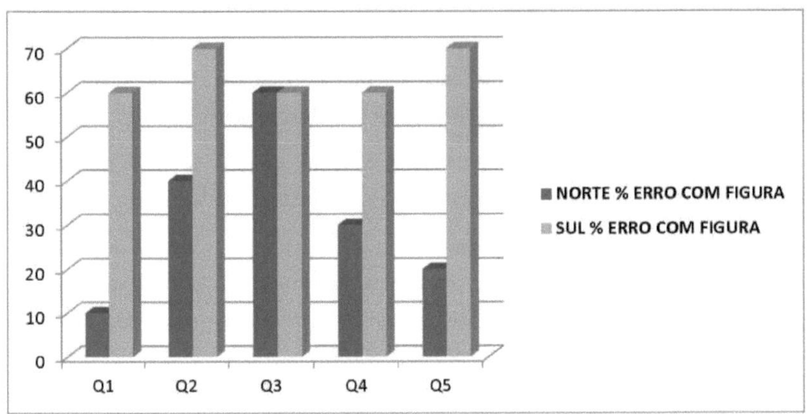

Source - the author

The object of study in mathematics is only accessible through its representation, so the more varied the ways of representing it, the greater the possibilities of understanding it. Duval further complements this idea by stating that the diversification of representations of the same object expands subjects' cognitive capacities as well as their mental representations. The development of the latter "takes place as an internalization of semiotic representations in the same way that mental images are an internalization of perceptions". (DUVAL, 2009, p.17).

7 **DATA CHECKS**

This paper examines the data collected during the research, so that at a later date a comparison of the data can be made, possibly with a new application of the tests. It's important to note that students are almost always only offered or presented with a single treatment, and rarely do they actually learn.

Graph 7 - Comparison of number of students and classes.

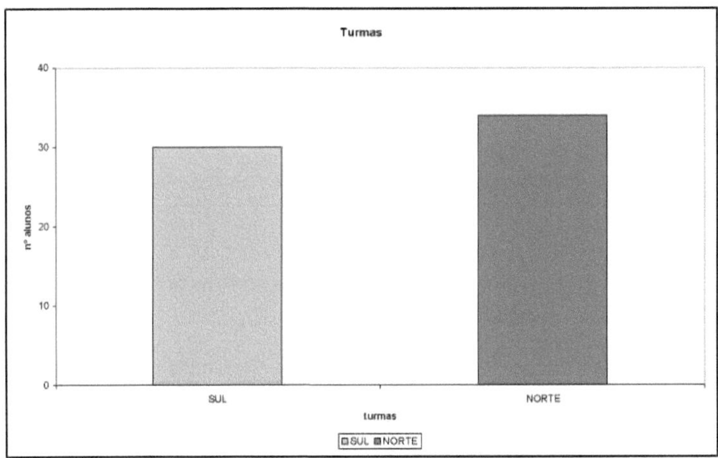

Source - the author

The graph? initially refers to classes that have been divided by region, one located in the north and the other in the south. The locations are very different, with approximately 32 km between them. The two classes were then divided into 67 students, but the southern class had a total of 32 students while the northern class had 35.

From now on, according to the legend, the north class will be identified as black and the south class as gray. The north and south classes have approximately 30 to 32 students each, but they were selected in advance because there was a need to hold classes on Saturdays, which made it impossible for some to attend because they help their families with their work, including at weekends.

When it comes to the question of gender, it can be seen that there are more females in the southern class, with 18 girls and 12 boys; while in the northern class there are 20 boys, but 14 girls.

Graph 8 - Comparison of number of students and gender.

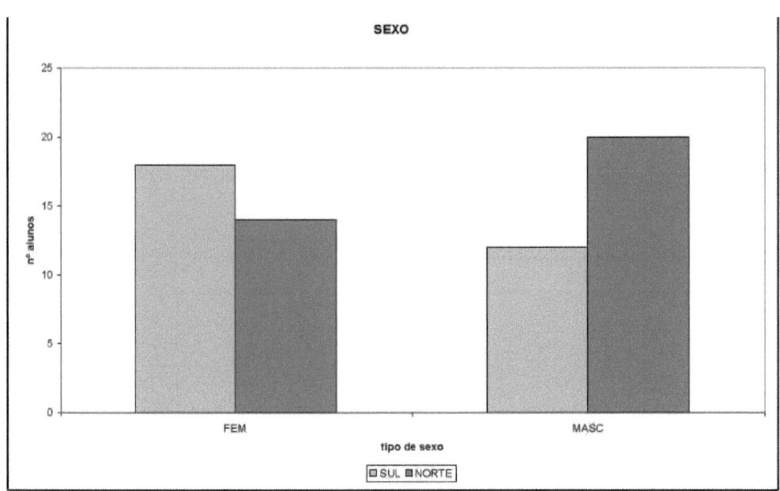

Source - the author

Age is undoubtedly a very important and fundamental variable in the analysis of knowledge, given that mental age for certain content can undoubtedly interfere with learning.

There were more mature students in the southern class, but younger ones in the northern class. At the end of the evaluations, we will record the degree of interference caused by this factor in relation to age, since it has a significant impact on the outcome of the teaching-learning process.

Graph 9 - Comparison of student numbers and age

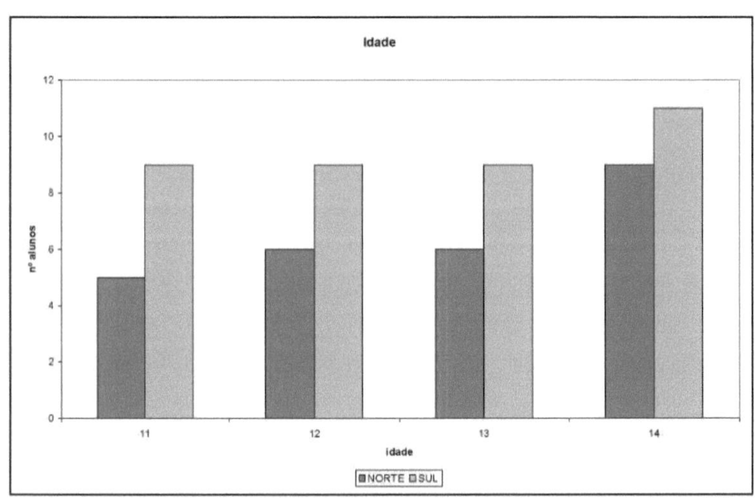

Source - the author

Graph 9 shows data on the students' liking for mathematics, where normally, when the subject is mathematics, there is a great deal of expectation as to how the students will react. In the course of the study, it became clear that the students used stigmas, almost always heard by the teacher, such as "I hate mathematics", "I don't know mathematics", etc.

This is why it is necessary to improve tools to change this behavior, where we will later have the opportunity to observe whether or not the inference is positive and the efficiency of the mechanisms applied during the study.

Graph 10 - Comparison: Number of students and liking for the subject.

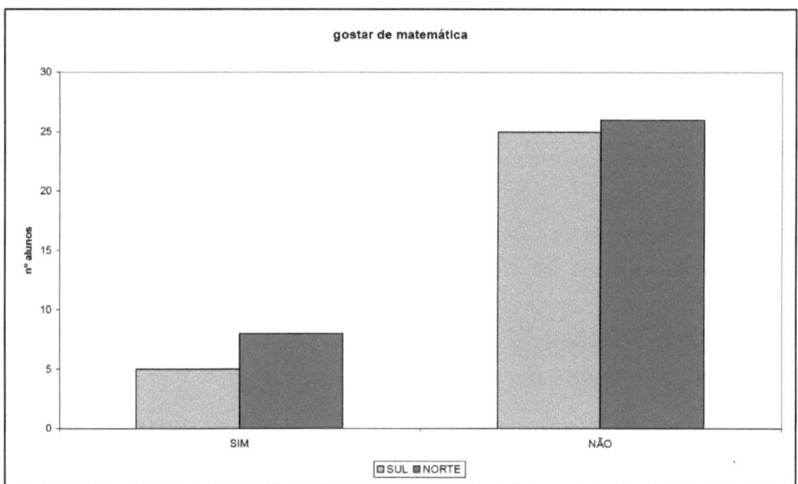

Source - the author

It can be seen that almost all the boys and girls involved in the research have difficulties in mathematics. This is a relevant reason to try to minimize the difficulties in understanding, thus reducing this fear, taboo or even lack of adequate and more consistent information to change the current configuration and thus provide teaching and learning with a high degree of satisfaction.

Graph 11 - Comparison between numbers of students and fractions.

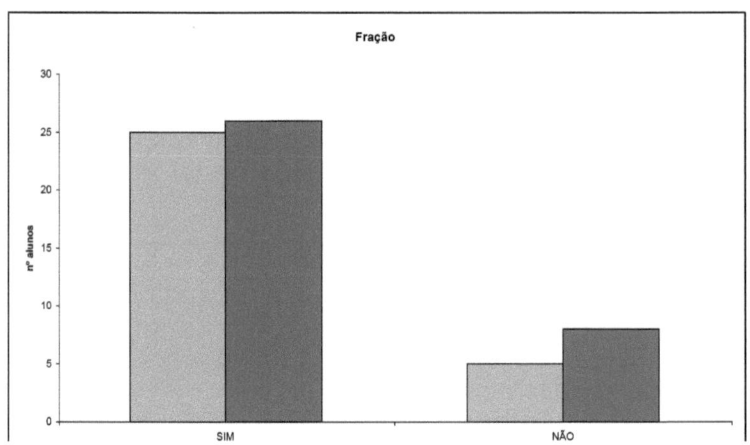

Source - the author

Graph 11 refers to the development of activities involving the content of fractions. It is clear that most of the students experienced the content, but it is necessary to assess what level of knowledge was acquired and in what way there was a positive contribution. Therefore, at the conclusion of the study, we will possibly arrive at the data gradually, so that we can assess whether or not interference is appropriate.

Graph 6 looks at the data in relation to the level of repetition. It's clear to see that there was almost no repetition, and the number of repetitions compared to one class and another is practically the same.

Graph 12 - Comparison of the number of students and the repetition rate.

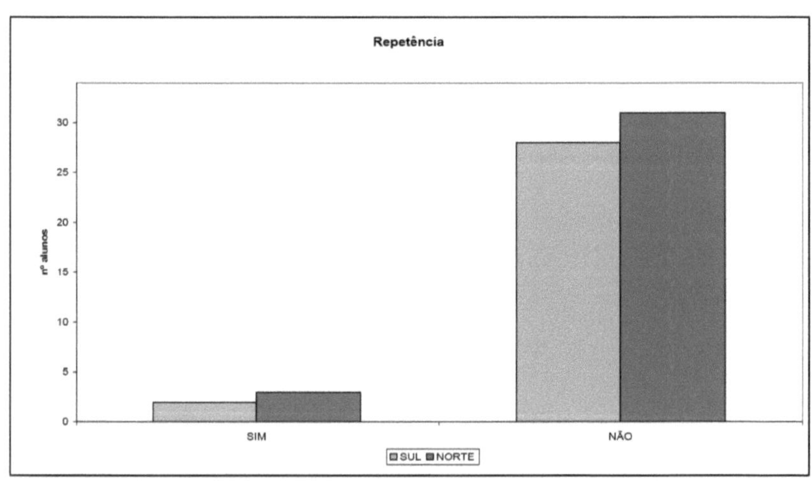

Source - the author

After collecting and analyzing the results, it became possible to draw up a profile of the two classes involved in the process, noting that there were no major factors that could possibly interfere with the development of the activities. We will then present the data by establishing the total number of successes and errors, which are shown in Appendix B.

8   DATA CHARACTERISTICS

First of all, a questionnaire (Appendix A) was applied, which included an approach to the students' characteristics, such as age, class, gender, whether they liked mathematics and, finally, the degree of repetition in the two classes.

Next up is the Fraction Representation Test (Appendix B), which consists of three questions. In question number 01 (one), a figure is presented, which refers to a rectangle; the students are asked what information can be transferred to the notebook with the help of the figure.

At this point, the aim is to observe the understanding of the language between the figure and the symbol; 03 (three) questions are then presented, referenced by the letters a, b and c. In letter a, we try to investigate by exploring the figure how many equal parts the rectangle has been divided into, with the aim of visualizing the figure.

Alternative b, in relation to the previous question, asks: Each of the parts represents what fraction of the rectangle? And finally, in the last question, letter c, the question is asked whether the painted part represents what fraction of the rectangle. After this, a critical analysis is carried out to check the level of prior knowledge about rationals.

Table 4 shows the number of correct answers between the northern and southern classes, in terms of percentage. It is shown graphically, demonstrating another form of language.

Table 4 - Number of correct answers between north and south classes by percentage.

Questions	Nº. Hits		Percentage n^0. Hits (%)	
1^a	10	15	33	44
2^a	11	16	37	47
3^a	10	15	33	44

Source: the author

Graph 13 shows the number of correct answers for the north and south classes, with a higher rate for the north class, with a favorable percentage of around 11%, which is quite considerable.

Graph 13 - Comparison of hits per class

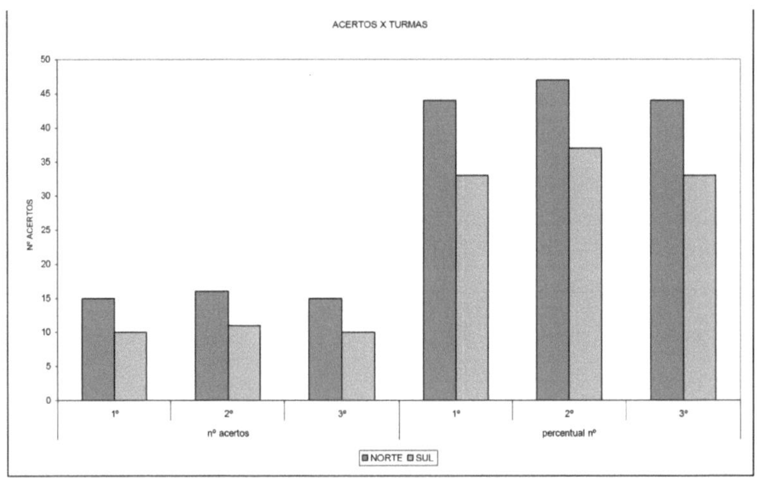

Source - the author

9 ANALYSIS OF THE CORRECT ANSWERS PER CLASS

The results of the evaluations and their data characteristics after the 2^a application are presented here. The test has been replicated, and some additional information has been added and will be included in Appendix (C), the information dealing with the content of fractions that is offered with the aim of stimulating the students, in view of the approaching end of the work.

The graphs that will effectively demonstrate the differences in the data match will henceforth be referred to as T1, T2 and DT1 and DT2. We now present the first sample after the second application in relation to the first.

Graph 14 - Difference in hits.

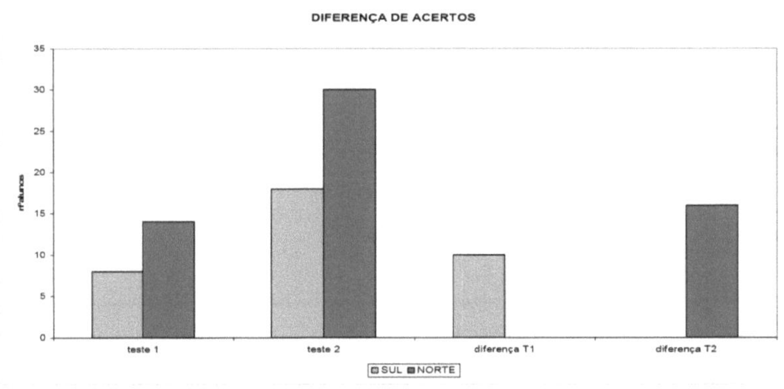

Source - the author

When analyzing this variable in relation to the first test and the second, it is clear from this sample that after the reinforcement of information and the commitment to the reinforcement classes. There was a very significant improvement, particularly in this sample, where it can be seen that the northern class developed more than the southern class.

The development of skills and, consequently, the participant's greater responsibility for learning, requires continuous practice, reflection, study and improvement. It is not possible to achieve this in one go, the degree of learning is variable for each individual, but it was clear in this sample that commitment gradually made a difference.

The following graph shows the difference in the number of correct answers in relation to age between the first and second tests in the class identified as the south.

Graph 15 - Hits in the south class x gender.

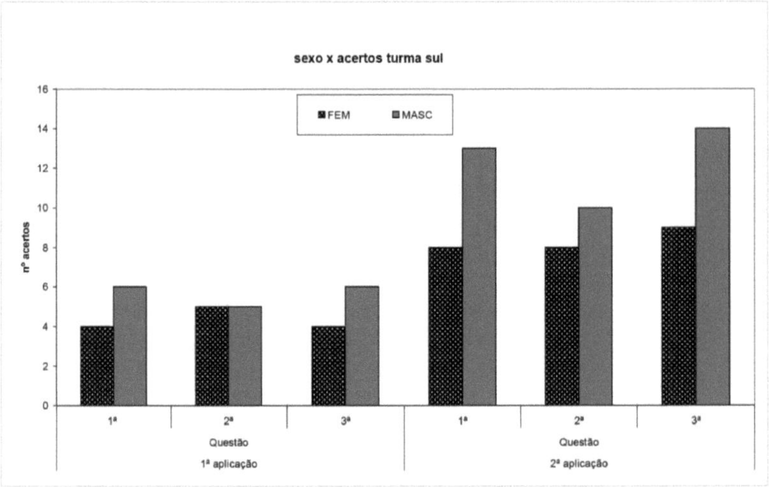

Source - the author

In this variable, it is clear that there has been a significant advance in relation to age, as the more adult students aged between 13 and 14 in the second assessment undoubtedly performed better.

If you look at the following graph, you can see that the northern class performed better than the more mature students, with a higher number of correct answers.

Graph 16 - number of correct answers x gender.

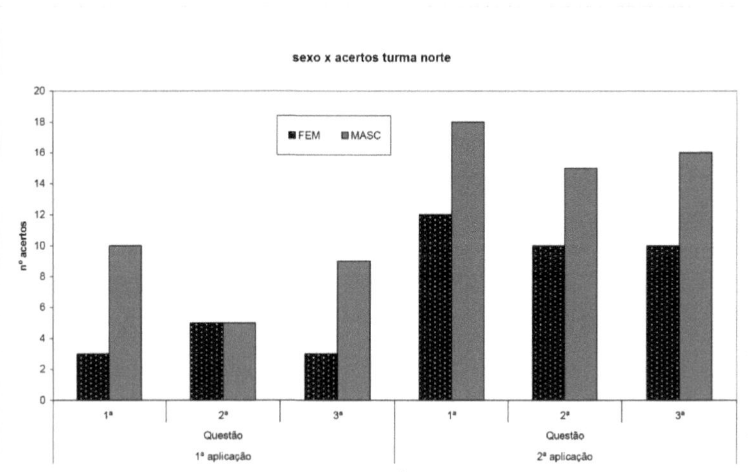

Source - the author

After analyzing the test data, it became clear that the northern class performed slightly better than the southern class. However, both classes improved considerably in the second application.

We believe that the way in which the lessons were conducted, through stimuli, the provision of concrete material, a new challenge. This has contributed to progress. In addition to the information on the semiotic representation system. As for the fact that the north class made more progress than the south class, this may be due to the location of the north class, which is further away from the center, although it is considered urban and not rural, but suffers less interference from the city.

9   **FINAL CONSIDERATIONS**

There are numerous difficulties and limitations encountered in the course of this research. Initially, it was clear that students in general, including those in the EJA, were focused on memorizing expressions and trying to learn by repeating examples. The students didn't have to mechanically memorize the description of the object, but learn its meaning in a broad way. Another relevant fact is undoubtedly the increase in reinforcement in the classroom, which requires careful handling. Despite innovations such as: games, games, etc. Another element that deserves to be highlighted is the fact that the teacher carries out the activities alone; the diversity of people may make it possible to achieve a more significant result.

The resources used to develop the activities are based almost entirely on Duval's theory, which expands on the idea of making students develop their cognitive skills through different forms of semiotic representation. From this perspective, the students didn't have to mechanically memorize the description of the object, but use it to develop arguments that help them understand its broader meaning.

The school needs to go beyond the mere transmission of information; it is important that the school values communication and can build a political-pedagogical project with the reality of the community in which people work. This new school must be attuned to the social networks that produce knowledge and can become organized citizens.

The EJA, in addition to proposing a continuous evaluation focused exclusively on young people and adults, necessarily needs to take into account that these subjects must be taken into account, that is, "what they think, where they come from, where they want to go", but the school that has been proposed to them is still the same as the one they attended, nothing has changed. There is a need for something new, something different, something attractive, something that motivates and stimulates. In this way, the school we want should be built together with this public, thinking along with them, with the community where they live, and this will certainly make it possible to achieve an efficient return on learning, promoting quality and inclusive education.

Our country is in a position to reorganize public education policies and establish the role that each citizen deserves. As long as this is not done in a conscious and consistent way, we will not evolve, because we will always be developing and our students will not play the simple role that is their right according to the constitution.

Students should always be encouraged to discover what is new, what they enjoy when they understand it, and then there is a paradigm shift in assessment, because students don't feel threatened or cornered as they do in traditional exams. This way, the results are certainly more significant, because the student feels protected, not punished as in the traditional model. Assessment should always be a continuous diagnosis in which the assessor has the time and patience to obtain the expected results. It is important for universities to rethink how to bring this

EJA audience into a new context, training specific professionals to meet this demand.

It is hoped that this study will not end now. It is hoped that countless people will be able to spread the knowledge and realize that the registers of semiotic representations are important for knowledge, not only in mathematics, but in various areas of knowledge.

10 **REFERENCES**

BRAZIL. Ministry of Education. **Leap into the future:** youth and adult education. Brasilia, 1999 (Série de estudos. Educaçâo à Distância, v.1θ).

CATTO, G. **Registers of representation and the rational number:** an approach in textbooks. 2000. 152 f. Dissertation (Master's Degree in Mathematics Education) - PUC-SP, Sâo Paulo, 2000. Available at: <

http://www.sapientia.pucsp.br/tde_arquivos/3/TDE-2007-07-04T07:24:43Z-3735/Publico/dissertacao_gloria_catto.pdf >. Accessed on: December 12, 2014.

DUVAL, Raymond. Registers of semiotic representations and the cognitive functioning of understanding in mathematics. IN: MACHADO, Silvia Dias Alcântara.

Learning in mathematics. 5. ed. Sâo Paulo: Papirus, 2009.

. **Semiosis and Human Thought:** semiotic registers and intellectual learning. São Paulo: Livraria da Fisica, 2009.

. **Écarts sémantiques et cohérence mathématique:** introduction aux problémes de congruence. Annales de didactique et de Sciences Cognitives, v1, IREM de Strasbourg, 1988. p. 7-25.

MACHADO, S.D. A. Aprendizagem em Matemàtica. 5 ed. Sâo Paulo: Papiros, 2009.

. **Educaçâo Matemàtica**: uma introduçâo. 2.ed. Sâo Paulo: EDUC, 2002.

FREIRE, Paulo. **Education and change.** Rio de Janeiro: Paz e Terra, 1989.

MARANHAO, M.C.; IGLIORI, S.B.C. Registers of representation and rational numbers. In: MACHADO, S. D. A. **Aprendizagem em matemàtica -** registros de representação semiótica. Sâo Paulo: Papirus, 2003. p.57-70.

RAMOS FARACO. Fractions without Mysteries. Sâo Paulo: Editora Atica, 2000.

SILVA, Z. A. Operations with Fractions in Their Various Meanings. Specialization Monograph. UFMS. Três Lagoas, 2002.

APPENDIX A - Questionnaire

Student profile.

Dear student, this test is designed to collect data for a monograph. Your contribution is essential.

You answer very carefully.

The$_v$ related questions from .01..3 04 mark only one= alternative as the answer.

1. What class are you in A ( ) B ( ). Gender. Male ($_w$) Female ( ).td=de ( ).

2. Do you like mathematics?

Yes( ) No( )

3. Do you use mathematics in your daily life?

Yes( ) No( )

4. Have you developed activities with fractions?

Yes ( ) No ( )

5. Are you a repeater?

Yes ( ) No ( )

Thanks for your cooperation!

APPENDIX B - Fraction representation test

1.   Look at the picture:

In this presentation, the student is asked what information they can transfer to their notebook with the help of the picture. The questions are then asked:

a)   How many equal parts has the rectangle been divided into?

b)   What fraction of the rectangle does each of these parts represent?

c)   What fraction of the rectangle is the painted part?

Once again, new figures are presented, but with comments on the previous illustration, at which point a connection is made between the symbol and the representation of the number. A moment later, we are offered notions of fractions, and various representations, saying how many parts the whole number has been divided into.

This is the time to present different situations to facilitate understanding, because understanding only happens when there is enough information.

APPENDIX C - Test 1

<u>Test it first;</u>

1 - The length of a plank is 20 m. How much is 3/5 of this board?

2 - If 2/3 of a road is 100 km, how long is it?

3 - I have 60 chips. My brother has 3/4 of that amount. How many chips does my brother have?

4 - A box contains 80 candies. How much do 2/5 of these sweets represent?

5 - You need 120 liters of water to fill 1/5 of a tank. What is the capacity of this tank?

Comments on the first test in the south zone:

1- Only 20% of the students managed to solve the first problem, even using a figure to help them find the result.

2- Only 10% of the students worked on the second challenge, because the problem required a higher level of interpretation, since the proposed problem starts by asking about the fractional part, blocking interpretation to a certain extent. Some students immediately before handing in the tests already mentioned comments about the difficulty of the second question, questioning the way it was interpreted.

3- In this third challenge, the results were more significant, with 30% of the students answering correctly, and the students feeling more confident in their interpretation.

4- This problem repeats the level of comprehension of the third, where the same 30% appears as a positive result, as the effort required to interpret is similar to the previous one.

5- In this fifth challenge, 10% of the students got it right, even though the difficulty level was similar to the first problem; it is believed that the margin of error was due to the student's lack of attention.

APPENDIX D - Test 2

1 - The length of a plank is 20 m. How much is 3/5 of this board?

Figure 3 - Length of a plank

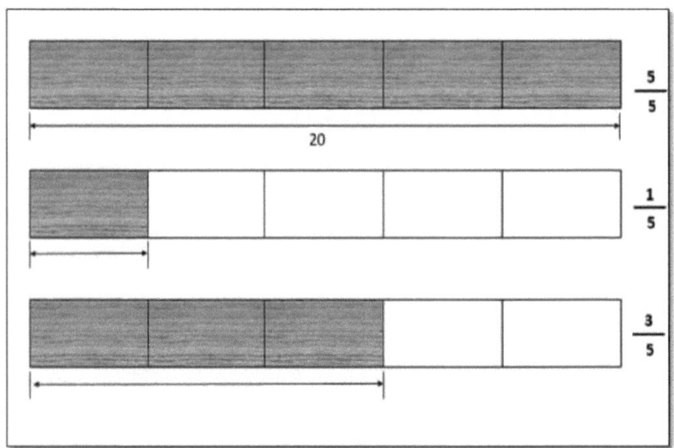

2 - If 2/3 of a road is 100 km, how long is it?

Figure 4 - Length of a road

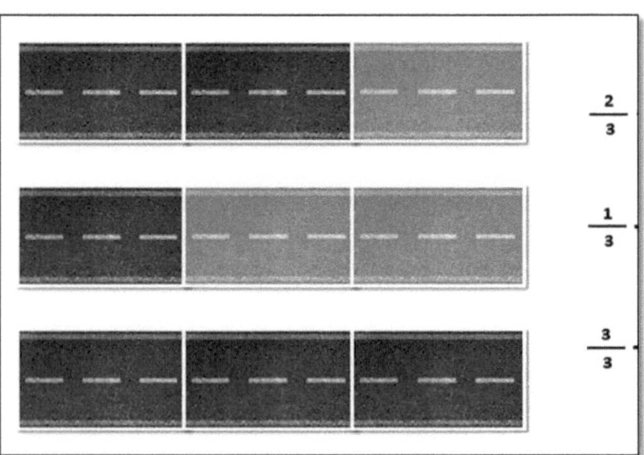

3 - I have 60 cardboard boxes. My brother has 3/4 of that amount. How many boxes does my brother have?

Figure 5- Boxes

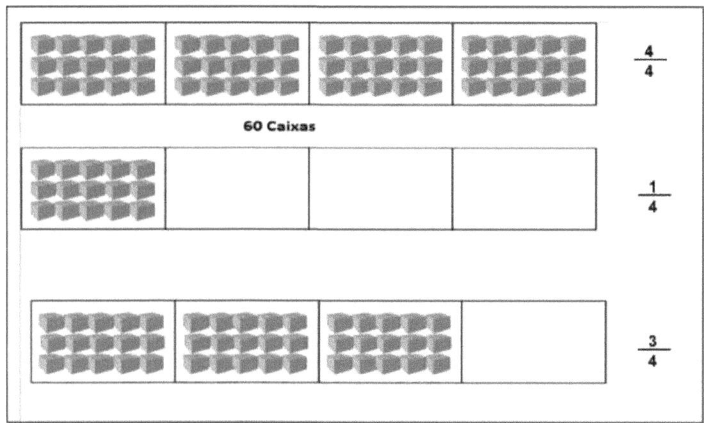

4 - A box contains 80 candies. How much do 2/5 of these sweets represent?

Figure 6- Candy boxes

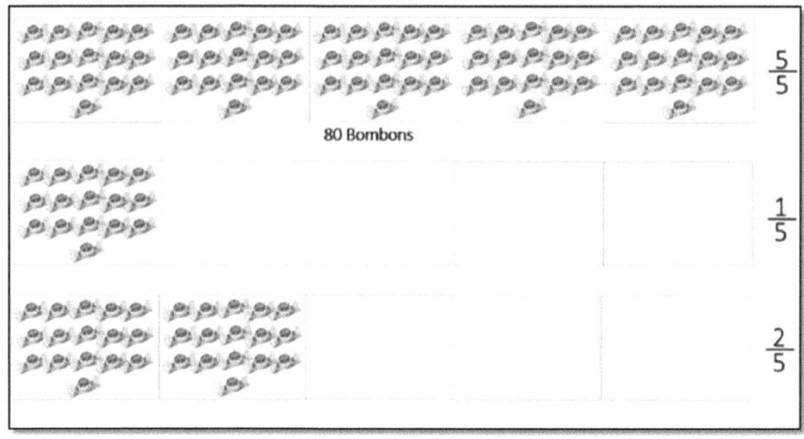

5 - You need 120 liters of water to fill 1/5 of a tank. What is the capacity of this tank?

Figure 7- Truck

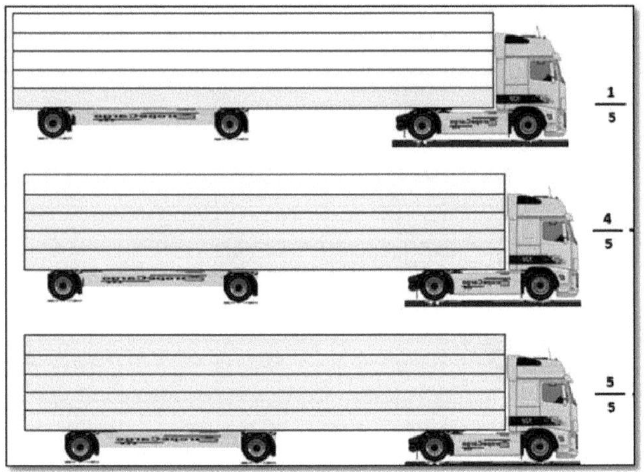

1
5

4
5

5
5

Printed by Books on Demand GmbH, Norderstedt / Germany